I0483262

Camille Goes Giant Bug Hunting

Robert D. Broughton, MS

Copyright © 2016 Robert D. Broughton, MS

All rights reserved.

ISBN: 10: 1523789263
ISBN-13: 978-1523789269

DEDICATION

Camille

Macro Photography By: Robert D. Broughton, MS

© 2016

All Rights Reserved No DUPLICATIONS of Photographs Permitted

Self-Published

ISBN: 10: 1523789263
ISBN-13: 978-1523789263

Special Thanks to the Cohutta Wilderness Area Tennessee/Georgia

ACKNOWLEDGMENTS

Cohutta Wilderness Area, Blue Ridge Mountains

Tennessee-Georgia

Blue Ridge Mountians.Com

Dobsonfly

8.3 Inch Wing Span found in Sichuan Province, China

Helicopter Damselfly
7.7 Inch wingspan found in South America

Giant Burrowing Cockroach
7 Inch in Length found in Queensland. Australia

Giant Camel Spider
Body 6 inches long Found in Iraq

Giant Walking Stick
7 Inches long found in North America

Giant Asian Hornet
2 Inches long found in Asia

Coreid Left-footed Bug
4 Inches long found in China

Giant Weta
4 Inch long found in New Zealand

Titan Beetle
7 Inches long found in Asia

Titan Beetle Face

Elephant Beetle
5 Inches long found in Asia

Hercules Beetle
5 Inches long found in Asia

Giant Metallic Ceiba Borer
4 Inches long found in Panama

Atlas Moth
10 Inch wingspan found in New Guinea

Giant African Millipede
15 Inches Long found in Africa

North America Giant Centipede
8 Inches long

South American Giant Centipede
12 Inches long

Saddle-backed Bush Cricket
3 Inches long found in Spain

Giant Scarab Beetle
4 Inches long found in Africa

Giant Click Beetle
4 Inches long found in the Amazon

Giant Long-Legged Katydid
6 Inches long found in Asia

Giant Isopod
14 Inches long found in Mexico

Tarantula Hawk Wasp
3 Inches long found in North and South America

Chines Pray Mantis
8 Inches long

Goliath Bird-Eater Spider
12 Inches long Found in South America

Weevil From Russian
3.5 Inches Long

Giant Spiny Stick Insect
5 Inches Long found in Australia

Polyphemus Moth
6 Inches long found in North America

Giant Stag Beetle
4 Inches long found in North and South America

Giant Campontus Gigas Ant
2 to 3 Inches long found in Oriental Region

Giant Bull Ant
1.5-2 Inches long found in Australia

Giant Bamboo Weevil
4 Inches long found in China

Rhinoceros Beetle
3 inches long found in Boreo

Robert Broughton is a school counselor in Northwest Georgia, he has written two books, :Krispy Kritter: A Book About Teasing and a novel called Ghost Tales of The State Line Mob. He is an semi-professional macro photographer with several published photographs. He uses his talents to teach children and share the wonderful world of nature with them. The series of books called Camille Goes Hunting- are adventures of discovering different insects and animals outside. Mr. Broughton lives in Cleveland, TN with his wife Donna and daughter Kaylie.

www.ingramcontent.com/pod-product-compliance
Lightning Source LLC
Chambersburg PA
CBHW050839180526
45159CB00004B/1965